Demóstenes Barbosa da Silva

Environmental Productivity of Energy Sources

Demóstenes Barbosa da Silva

Environmental Productivity of Energy Sources

Metrics for the sustainability of energy generation

ScienciaScripts

Imprint
Any brand names and product names mentioned in this book are subject to trademark, brand or patent protection and are trademarks or registered trademarks of their respective holders. The use of brand names, product names, common names, trade names, product descriptions etc. even without a particular marking in this work is in no way to be construed to mean that such names may be regarded as unrestricted in respect of trademark and brand protection legislation and could thus be used by anyone.

Cover image: www.ingimage.com

This book is a translation from the original published under ISBN 978-3-330-76265-7.

Publisher:
Sciencia Scripts
is a trademark of
Dodo Books Indian Ocean Ltd. and OmniScriptum S.R.L publishing group

120 High Road, East Finchley, London, N2 9ED, United Kingdom
Str. Armeneasca 28/1, office 1, Chisinau MD-2012, Republic of Moldova, Europe
Managing Directors: Ieva Konstantinova, Victoria Ursu
info@omniscriptum.com

Printed at: see last page
ISBN: 978-620-8-38673-3

I dedicate this thesis to humanity.

Thank you

This study was carried out at the Graduate Programme in Energy at the Institute of Energy and Environment, University of São Paulo, from July 2011 to January 2016.
I would like to express my sincere appreciation to my supervisor, Professor José Goldemberg, for the productive dialogues we had about this study. His guidance for my reflections, his support and encouragement, always present in our dialogues, and his various observations were invaluable to me throughout the time I dedicated to developing this study.
I would also like to acknowledge the encouragement received from Prof Dr Ildo Sauer, Director of the Energy and Environment Institute, for his complicity in the efforts to develop this study. My thanks also go to the Brazilian Institute for the Environment and Renewable Natural Resources (IBAMA) and the National Electric Energy Agency (ANEEL), which provided me with essential data and information on environmental impact assessment and its correlation with the technical characteristics of power generation facilities.
I extend my appreciation to the Professors and colleagues at the Institute of Energy and Environment for the precious moments of learning and encouragement.
Finally, to my family, I probably haven't thanked them enough for all the time I gave up to do this study.
There are other people to whom I don't express my gratitude specifically in this document, but to them I owe a debt of gratitude.

The development of Western science is based on two major achievements: the invention of the formal logical system (in Euclidean geometry) by the Greek philosophers, and the discovery of the possibility of discovering causal relationships by systematic experiment (during the Renaissance).

Albert Einstein (1953)

SUMMARY

There is a fundamental relationship between public policies and productivity (Georgescu-Roetgen's, 1976). "Environmental productivity of an energy source" is defined as the relationship between the energy supplied by a given source and the extent of the atmosphere and biosphere used in its production. It is a practical indicator, formulated on the basis of parameters commonly used in decisions on new sources of electricity generation for the National Interconnected System (SIN) in Brazil. This indicator can help public policy makers establish minimum requirements and incentives for energy sources that produce lower environmental impacts. Calculations show a significant increase in the "environmental productivity" of hydroelectricity and sugar cane, and relative stability in the case of oil. Its application to various energy sources in the SIN indicates a vertiginous increase in environmental productivity with the introduction of wind power into the Brazilian energy matrix, and an even greater increase is expected with the introduction of solar power. The "environmental productivity of energy sources" in a global perspective can help the planet in a virtuous transition of the global energy matrix from its current composition, with a preponderance of conventional fossil energy sources, to renewable energy systems, if policymakers can align energy policies with minimum "environmental productivity" requirements. Appropriate priorities and transitions must be incentivised to enable countries to replace fossil fuel-generated energy with renewable energy sources (Wang, 2011). It is important to help decision-makers and energy users understand the general sense of "ecological footprints" and the planet's limits in regenerating ecosystems. In order to reduce pressure on the atmosphere and biosphere, it is also essential that public energy prices reflect the real costs incurred. The reason why the "environmental productivity of energy sources" can be used to reduce pressure on the atmosphere and the biosphere is the comparability between sources, which can be established from the methodology of its calculation, applied to a set of sources, or alternative sources. Decision-making processes for implementing, or even decommissioning, energy sources are generally based on sequential ordering of energy sources over time, or on alternative sources capable of meeting energy supply needs, according to

their costs, the needs of the power system to which they are connected, and assessments of the quality and magnitude of their environmental impacts. However, environmental impact assessments consist of studies whose terms of reference and results explain local impacts, and support individualised environmental licensing decisions that are not directly comparable with each other or with common general requirements. Finally, it is considered that there are no negative impacts associated with policies that incentivise the use of renewable energies for energy users, and this makes it legitimate to create a stimulating environment for energy market agents in a continuous search for greater environmental productivity in the use of energy sources.

Keywords: *Environmental Productivity, Energy Policy, Renewable Energy, Technological Transition of Energy Systems*

SUMMARY

CHAPTER 1

Introduction

1.1 Background to the study

The energy sector is structurally complex: capital-intensive, the decision-making processes for investments, implementation and operation of energy generation facilities require planning years in advance to meet anticipated demands; it is a sector that has historically been subject to strong political and economic interests, particularly in developing countries, as well as being increasingly vulnerable to unpredictable and sometimes uncontrollable circumstances of some factors, such as climatic conditions and political changes, and economic and social growth (Khayyat, 2015).

This complexity and these vulnerabilities have motivated interventions from various segments of society; academia, where numerous contributions advocate technological, organisational, regulatory and public policy improvements, as well as diversifying the use of available energy sources that are not yet sufficiently used, notably biomass and solar irradiation in developing countries; the energy industry, which through various trade associations, plays a leading role in diagnosing, criticising and complaining about energy policy and regulations; the media, which have specialised in carrying out critical analyses of the facts of the energy sector and the implications of public policies on them, while reporting on the facts of the energy sector for society; parliaments, where bills promoting changes to the legal and regulatory framework of the energy sector are discussed; and governments, which face major challenges in securing energy supplies, acting on the markets by means of incentive and limitation instruments, which should ideally be anticipated in energy policies and established in law (Goldemberg, 2012).

The debate is in fact broader, a debate about the limits of the planet beyond which the continued use of some natural resources, as they are currently being used, puts the planet's capacity for regeneration and the sustainability of the current development model at risk. It has motivated the engagement of various researchers and civil society organisations, and is expanding with the support of citizens, companies and, although less evident, governments. Measuring the speed with which humanity uses the available natural resources over a given period of time gives an indication of how many planets are needed each year to meet human needs. The calculation of this measure, based on data on the flow of natural resources being used in 150 countries in both the western and eastern hemispheres of the Earth, resulted in an indication that humanity's ecological footprint has already exceeded the Earth's regenerative capacity by approximately 44 per cent (Wackemagel and Rees, 1996).

But in order to better understand the question of the optimal use of natural resources, questions arise about how these resources are used, and how, with what technologies, the same quantities of goods and services demanded could be obtained, while at the same time using fewer natural resources. In other words, how to ensure the greatest productivity in the utilisation of these natural resources. The answers to these questions have not yet been formally proposed. They will probably require specific metrics that make it possible to incorporate mechanisms for increasing productivity in the use of natural resources into decision-making processes.

1.2 Objective of the Thesis

The aim of this thesis is to clarify how the concept of "environmental productivity of energy sources" can be used to reduce unnecessary environmental pressures on the biosphere and atmosphere.

1.3 Proposed questions

In order to achieve the above objective, the following questions were considered, taking the case of Brazil as a basis for analysis:

1. What does the current energy matrix look like and what trends can we see in its evolution, in terms of environmental pressures resulting from energy generation?
2) Considering the energy production sources and technologies currently available, how can we envisage the evolution of environmental pressures resulting from their use to meet the country's energy demand?
3) How is the environmental productivity of energy sources in Brazil analysed from an individual and energy matrix perspective?
4. How can energy policy incorporate the concept of the environmental productivity of energy sources in order to reduce the environmental pressures arising from electricity generation?

1.4Scope and delimitations

The establishment of the concept of "environmental productivity of energy sources", an example of its calculation and recommendations for its use as an element of energy policy, constitute the fundamental scope of this thesis.

The definition of the concept of environmental productivity is based on the foundations of the theories of production described by the thinkers of classical economics and results in a mathematical formulation for the use of known variables in energy generation projects and aggregates of energy sources in an energy matrix.

To illustrate the use of the concept of environmental productivity of energy sources, we used data on energy sources kept on file by the National Electric Energy Agency (ANEEL) and the Brazilian

Institute for the Environment and Renewable Natural Resources (IBAMA), relating to the various energy generation facilities, and also aggregate energy production data from hydroelectric, sugar cane and oil sources, obtained from official publications by the Ministry of Mines and Energy, with which the calculation of the "environmental productivity" of aggregates of the Brazilian energy matrix was tested. This test was carried out to illustrate and evaluate the results of applying the concept of "environmental productivity of energy sources".

The concept of the "environmental productivity of energy sources" is used to develop analyses on how to increase environmental productivity, highlighting the greater production of energy from sources with a low environmental impact, such as renewable sources, and the possible replacement of conventional energy sources with renewable sources with a low environmental impact, as a fundamental means of increasing the "environmental productivity of energy sources".

With this fundamental perspective, the discussion on renewable energy sources that can help increase the environmental productivity of the energy matrix is developing.

On the other hand, increasing the efficiency with which energy is used has proven to be a reliable way of promoting energy savings and, it can be said, is also a coadjuvant in promoting greater "environmental productivity" in the energy matrix, insofar as, for a given level of end-user needs, the demand for energy will be lower and, consequently, the need to use natural resources will also be lower.

From this other perspective of the supporting role of energy efficiency, considering the direct cause and effect relationship of energy consumption by end users, reflections have also been developed on the energy consumption of some classes of end users of energy generated and commercialised in Brazil, and also on how energy is generated (Goldemberg, 1995).

Thus, the concept and examples of its application, the alignment of the metric with elements of an energy policy, such as regulatory and fiscal incentives for expanding the supply of energy from renewable sources, and for increasing efficiency in energy consumption, are developed in this Thesis without going beyond the limits of the discussion on how to use the concept of "environmental productivity of energy sources" in energy policy to reduce unnecessary environmental pressures on the biosphere and the atmosphere.

1.5 Thesis structure

The thesis presented here is made up of six parts, organised into sections in such a way that the discussion and analyses presented in each one conceptually form a well-characterised section or aspect of the reasoning. The first section is the introduction, which presents the background to the subject under study, the aim of the thesis, the reasoning behind the study and an explanation of its importance, and finally the scope and structure of the thesis.

The second section is a review of the literature on the topics discussed in the thesis and their relevance to the study. Developments made in previous works that have significance for the present study are commented on.

The third section explains the concept and method of calculating the "environmental productivity of energy sources". A conceptual discussion of "ecological footprint" is presented, and comments are made on the applicability of the concept of "environmental productivity of energy sources" as a practical way of guiding efforts and decisions to reduce the "ecological footprint" of energy use. Some conclusions on how the concept of productivity could be associated with regulatory incentives for energy efficiency are drawn from an analysis of the industrial energy segment, which is the largest among the classes of energy users. This analysis, although interesting for helping to understand the effect of applying concepts such as energy efficiency and environmental productivity, is not essential and has been included in Annex I.

In the fourth section, the results of the calculations of the environmental productivity of energy sources are examined and organised in scientific categorisation.

The fifth section presents a discussion of the results obtained by applying the concept of "environmental productivity of energy sources" to the generation plants that make up the SIN in Brazil.

The sixth section draws attention to some proposals that may be useful to energy policymakers and anyone interested in exploring the concept of "environmental productivity of energy sources" or using it in specific applications.

The seventh section presents recommendations on the use of the concept of "environmental productivity of energy sources" as an adjunct to the adoption of energy policies geared towards the transition of energy matrices towards compositions with lower environmental impacts.

The last section presents the bibliographical references and their authors.

CHAPTER 2

Literature Review

The literature selected reflects recent work relevant to the research carried out. Although the specific topic of the environmental productivity of energy sources has not been widely published, the literature consulted provides a clear overview of the scope of the study, as well as inspiring an in-depth understanding of the field of study as illustrated in the following sub-sections.

2.1 Energy-environment causal link, "ecological footprint" and energy policy

Energy and other policies can be aided in their implementation by analytical tools and metrics to describe and provide parameters for decision-making based on complex interactions between environmental, energy, water, and other social systems (Galli et al. 2012; McGlade et al. 2012; Pfister, Koehler, and Hellweg 2009; Boulay, Hoekstra, and Vionnet 2013). The causal link between energy production and consumption, warming of the Earth's surface and biodiversity loss has drawn attention to the interdependence of energy, environmental and social development policies for low-income socio-economic classes.

Although it is clear and increasingly recognised that there are global crises that require articulated public policies, such as energy and water (Srinivasan et al. 2012; UNDP 2006), and climate change threatens the viability of these policies (Gleick 2010; Võrõsmarty et al. 2010; Oki and Kanae 2006), public policies and their formulators have failed to consider this causal nexus and its implications. Questions have been raised in several countries about how energy policies can mitigate climate change and simultaneously guide the adaptation of energy generation systems to climate change, particularly with regard to hydrological uncertainty (Fulton and Cooley 2015).

Some metric approaches to the limits that ecosystems can withstand have been used to motivate public policy makers and governments to adopt more rigorous and assertive measures to limit the impacts of anthropogenic activities on these systems. These approaches have been recognised as expressions of the "ecological footprint" of a given anthropogenic activity or set of activities. The term "ecological footprint" was introduced in the early 1990s by Mathis Wackernagel and William Rees. It was in fact one of the first attempts to measure the planet's capacity to support all anthropogenic activities. The authors adopted the concept of how many planets would be needed to support human activity with the resources required from nature in a given year. In 2008, they made an estimate of humanity's "ecological footprint" based on data obtained from 150 nations on the use of natural resources since 1980. They concluded that humanity's ecological footprint has exceeded the planet's regeneration capacity, and in 2008 the capacity of the Earth's ecosystems was exceeded by approximately 44 per cent.

The economic and financial analyses that underpin investment decisions today are still based on a vision of linear cause and effect interactions in the various systems of the physical environment. This view, that anthropogenic action on matter results in orderly, known and predictable effects, is present both in the decision-making processes for implementing new energy generating sources, which are characterised as microeconomic decisions, and in energy policies whose effects are macroeconomic. The practical effect of this view on investment project decision-making processes is that projections of economic costs and revenues over time, and discounted cash flows, generally favour the recovery of short-term values, without considering the costs and benefits associated with environmental assets, which are generally appropriated over the long term. On the other hand, there are propositions in the literature that current economic thinking should be based on the Laws of Thermodynamics, with due consideration for Biodiversity and the limits of the respective biological systems (Wijeratne A.T., 2015). According to these propositions, anthropogenic activities invariably tend to increase the entropy of the systems on which they act (Styer D. F., 2000), and therefore the economic equilibrium of these systems, or of markets associated with them, is an apparent phenomenon, insofar as their continued growth will result in unforeseen imbalances arising from the inability to absorb the effects of growing entropy.

These propositions underpin a new vision of the nature of the economy, which has been characterised as the bio-physical nature of the economy.

There is a question, the answer to which does not yet seem to be evident and sufficiently accessible in the literature for a significant part of society, especially for energy policy makers and decision makers of investment projects in energy sources, which is the question of how the expressions of the "ecological footprint" can be taken into account in governance and public policy making decisions, and particularly in decisions to install new energy sources. This was considered a starting point for a possible contribution to the current debate with the results of this thesis.

2.2 Efficient use of natural resources

Resources around the world are considered measurable and quantifiable. This inference about resources around the world is always associated with a certain period of time (Williams, McKane, Perry, Aixian & Tienan, 2005). It is therefore essential to ensure that waste in the use of natural resources is reduced in order to preserve the planet from resource scarcity. The reason for this idea is quite consistent with the logic of resource utilisation. The efficient utilisation of resources, for example minerals, energy and water, raw materials and so on, requires sustainability (Goldemberg, 1995).

Resource efficiency can be characterised as the use of natural resources in the best way, as would normally be prudent, minimising the effects of their use on nature (Smil, 2008). Resource efficiency is not just a natural concern; it is also an activity that generates value by increasing productivity and reducing costs for users, for example, raw material costs, energy costs and operating costs (Williams, McKane, Perry, Aixian & Tienan, 2005). Similarly, from a business point of view, productivity contributes to the security of resource supply by ensuring that customer demand can be met in the course of business (Borg, 2008). By guaranteeing productivity in the use of resources, benefits are also ensured for the economy, increasing profits for businesses and reducing expenses for end users. Furthermore, the ability to use resources more productively can increase profitability without additional environmental impacts, and from an ecological point of view, greater efficiency in the use of resources reduces greenhouse gas emissions, reduces the use of natural resources such as water and the physical environments of the biosphere and atmosphere, ensuring better use of raw materials (Khayyat, n.d).

Efficiency in the use of resources is, in terms of value for production, equivalent to the acquisition of raw materials. The more efficiently resources are utilised, the less waste is generated. The reuse of materials and the elimination of duplicate procedures result in increased efficiency in the utilisation of resources (Goldemberg, 1995).

All over the world, energy, whether renewable or not, is a resource and so the fact of ensuring efficiency in the use of energy is already halfway towards increasing environmental productivity, in the terms of the approach adopted in this work. Energy efficiency is one of the practices that can secure resources in various production lines, and is extremely effective when put into practice as part of sustainability policies (Borg, 2008).

2.3 Energy efficiency, an adjunct to environmental productivity in an energy matrix

Energy efficiency can be characterised as using less energy to provide the same amount of a product. This definition can be linked to the proportion of energy in the energy supply in any given process. In practical terms, the least amount of energy that would be sufficient to produce a given good in a process (Simón, 2012).

Based on the same rationale, it can be said that greater efficiency in the use of energy will result in greater environmental productivity of an energy matrix, since smaller quantities of the biosphere and atmosphere will be used for the same utility of energy.

The industrial user segment accounts for the largest share of total energy consumption of all consumer classes. It is in this segment that rationality and efficiency measures in energy use show the most significant results, without disregarding the importance and potential contribution that the other segments make to improving the economy's overall productivity.

Considering the main motivations for the practice of energy efficiency in companies, particularly industrial ones, three main connections with productivity in general can be seen, involving the economic and environmental senses:

- Rising energy costs,
- New environmental regulations that induce the reduction of greenhouse gas emissions (mainly CO_2), for example the 1997 Kyoto convention, the 2009 Copenhagen Accord, and the programmes expected to be developed from the decisions of the Conference of the Parties to the Global Climate Convention in Paris in 2015, and
- Subsequent effects of end users changing their consumption decisions to more sustainable options and companies seeking to bring their products and image closer to the context of greater sustainability.

In any case, among these three main motivations, in a related study, the increase in energy costs is the main motivating factor. This is attributed to the fact that the world wars and also the local wars that are still taking place on the planet caused and have caused expected increases in energy costs and different variables today have contributed to the persistence of this increase. This has drawn the attention of many energy experts to the rationality of measures that can guarantee the sustainable use of energy and its efficiency. It has also drawn attention to the issue of rising energy costs for small-scale companies in Brazil. It is widely recognised in Brazil that an important part of the value gains from increased productivity come from greater energy efficiency. A number of other pressure factors on small businesses, some of which are more important in the short term, overlap with issues of greater efficiency in the use of environmental resources (Pearce, 1991).

The benefits associated with promoting the productivity of energy generation and use, both from the point of view of energy consumption and also from the point of view of its production, which involves both economic and environmental productivity, are undoubtedly of high value to society, materialised in two results. The first is better financial performance, with lower costs and higher returns; the second is environmental sustainability. Better financial performance, with lower costs and higher returns on investment, is particularly important for small companies, especially in Brazil. Focusing on the financial advantages of greater productivity contributes decisively to practical and more sustainable development (Simón, 2012). (Cappers and Goldman, 2010), however, see this change of context with policies that promote greater productivity, as well as incentives for energy autonomy, as an additional advantage. Energy autonomy must be considered an imperative element by energy users who depend on large amounts of energy.

Nevertheless, some difficulties stand in the way of their implementation (Rohdin and Thollander, 2005) and these obstacles are clarified by the hypothesis that there is a certain resistance to innovations. This resistance to innovations is highlighted by arguments generally used to the effect that not all innovations have been sufficiently tested, and the financial areas of companies take this into account in their positioning, as well as the scarcity of data to justify the economic effectiveness of innovations; as a consequence, several opportunities to generate value are missed; this behaviour results in low interest on the part of administrations in energy efficiency and productivity projects. However, to help overcome these difficulties there are various possibilities for energy efficiency projects and programmes (Cost et al, 2007), which can help both energy end users (energy buyers) and energy suppliers to make progress in economic and environmental efficiency and productivity.

2.4 The environmental impacts of energy production and use

The environmental impacts of energy production and use arise from the economic activities of energy production and use (Ec.europa.eu, n.d). Health sciences and environmental sciences have established important insights into the connections between energy production and utilisation and their respective environmental impacts. Studies show that habitats are affected, exploitation beyond the limits of natural regeneration of renewable resources, emissions from energy production and use, and climate change are among the most crucial environmental issues. Studies have detailed and evaluated the loss of biodiversity and health problems. Unnecessary environmental pressures are related to the extraction and transformation of materials, as well as energy production and its sources. Some scholars have paid close attention to the causes of these pressures that are exerted on the environment and result in its depletion. Other studies have repeatedly recognised that the use of fossil fuels is the energy sector's biggest problem.

This school of thought provides an insight into the problem of energy production and use, a description of it and an analysis of it. However, it does not offer the resources needed to implement a solution, to stop the growth of these issues, nor does it pay much attention to the magnitude of the help it can give. It gives a clear view of the knowledge gap so that future studies can go deeper into this topic by scholars, researchers and scientific communities.

Environmental issues in a global perspective are like after-effects associated with certain components that represent recognised dangers. And the greater or lesser intensities of human activities become the main consideration of this danger. This is a direct result of the expansion of the world's population, energy consumption, level of physical activity and so on (Dincer, 1999). More and more often experts have focused on the natural effect of emissions of SO_2 , NO_X CO_2 and other gases. Today, energy utilisation has become one of the fundamental elements that is not disregarded when it comes to the issue of technological advancement, which in turn has been motivated by basic perceptions such as "improvements that meet the needs of present society without compromising the ability to meet the needs of future societies". Energy supply is one of the elements whose current practice can contribute enormously to the sustainability of the future, and renewable energy sources are exceptionally encouraging as parts of the solution (Dincer and Rosen, 1998).

This sense is very important for users in general and for companies in particular, due to their contribution to total energy consumption, in that it disseminates knowledge about the use of renewable energy sources, such as wind, solar and other renewable energy sources, the utilisation of which is becoming viable as a result of technological innovations. These sources are exceptionally encouraging to believe that a sustainable future in energy production will be feasible as a result of the simple use of equipment development and manufacturing arrangements to harness renewable energy sources (Pearce, 1991).

Population expansion and the consequent growth in activities will result in an increase in energy production and use; and because the world's energy matrix is contingent on energy sources that cause a large part of the environmental impacts on the atmosphere and the biosphere, this has provoked intense discussions about the ecological effects related to energy consumption, for example, the precipitation of corrosives, the depletion of ozone in the stratosphere, climate change, etc.

Acid rain

The result of the emission and mixing of wet and dry materials from the environment that have abundant quantities of nitric oxides ***(NOj*** and sulphuric anhydride (SO_2). It is also generated from the burning of fossil fuels, for example in non-ferrous metal foundries, transport vehicles and so on. Processes related to the production and use of energy are also notable sources of acid rain, and in this respect, countries in which energy generation from fossil fuels is at the top of their energy mix represent veritable deluges of acid rain (Elbehri, Segerstedt & Liu, n.d). Technological advances have been made in the production of fuels that are still fossil fuels, but with a lower composition of polluting elements, as is the case with S10 diesel in Brazil, whose composition has a reduced share of sulphur, and also in engines that use fossil fuels, with the introduction of equipment that retains part of the pollutant emissions in the exhaust gases.

But these technological advances are not enough to reduce the serious problem posed by the use of fossil-fuelled vehicles.

Stratospheric ozone depletion

The ozone present in the stratosphere plays a notable role in retaining ultraviolet (UV) and infrared radiation. The depletion of the stratospheric ozone layer, which is known to be related to emissions of chlorine fluoride carbides ***(CFCs)*** and nitric oxides ***(***NO_X ***),*** is a major ecological issue (Heijman, 1990).

As a result of the depletion of the ozone layer, ultraviolet (UV) and infrared radiation cause skin and eye diseases.

Energy production and utilisation processes are a small part of the problem of stratospheric ozone depletion; in fact, the emissions of CFCs used in refrigerators and air conditioning are the villain, but they add to the environmental impact by emitting nitric oxides.

Climate change

Climate change is the after-effect of multiple small impacts on the interface between water vapour and fog in the atmosphere, causing the planet's temperature to rise (Dinda, n.d). In addition to CO_2 whose increased concentration in the atmosphere accounts for around half of the problem of climate change, there are other gases, such as *CH^*, *CFCs*, N_2O, and SO_2 that contribute to worsening impacts on the atmosphere (McCarthy, 2001).

These gases are commonly emitted as a result of current anthropogenic activities, some of which are related to the production and use of energy.

These activities can be altered by changing the way in which energy is produced and used, for example by adopting more efficient energy production and utilisation practices, and by using renewable energy sources that do not emit such gases.

Table 1 illustrates all these gases, their related effects and the relative intensity of the environmental impact of each pollutant.

Table 1. Gas pollutants, environmental impacts on the atmosphere and relative intensities

Pollutant	Greenhouse effect	Stratospheric ozone depletion	Acid rain	Smoke
Carbon dioxide CO_2	Intense impact	Medium impact		
Methane CH_4	Intense impact	Medium impact		
Nitric oxide *NO* e Nitrogen dioxide NO_2		Medium impact	Intense impact	Intense impact
Sulphur dioxide SO_2	Reduced impact	Intense impact		
Chlorofluorocarbons *CFC*	Intense impact	Intense impact		

Source: McCarthy, 2001

Charcoal plants

Coal is still the energy source individually responsible for most of the electricity generated in the world, accounting for around 40 per cent of the total electricity generated in the world in 2015 (International Energy Agency, 2015). Various technological developments have been made in the direction of reducing the environmental impacts of these plants. Although the negative impacts of these plants cannot be eliminated, they could be reduced with the introduction of gasification technologies, pressurised fluidised beds and carbon dioxide scrubbers (Ekins, Pollitt, Summerton & Chewpreecha, 2012).

Photovoltaic (PV) panels

Photovoltaic (PV) panels have a very low risk of environmental impact and can be widely used in tropical areas where there is an abundance of exposure to solar irradiation, with a low risk of environmental impact.

Productivity has become progressively competitive compared to other sources. The problem related to the high cost of this technology can be minimised if the scale of its use is expanded sufficiently, which can be done by enabling end users to use available areas on the roofs of their facilities to install photovoltaic panels.

Hydroelectric power stations

Hydroelectric power plants have a significant impact on biodiversity, as shown in various studies, in that they drastically alter land use in areas of great importance for the preservation of terrestrial and aquatic biodiversity. And although difficult to measure, greenhouse gas emissions from the reservoirs of hydroelectric power plants are recognised as contributing to climate change, especially in Brazil. Recent studies have shown that in the past there were insufficient standards for the safe planning and design of hydroelectric plants. One issue that is present in the growing controversy that arises when a large hydroelectric plant project is made public is the magnitude of the impacts and the determination of the correct number of people who should receive compensation for the damage caused by the implementation of these hydroelectric plants. Even with these uncertainties, hydroelectricity represents an effective alternative for reducing environmental impacts in electricity generation, particularly in developing countries, and with limitations on the size of the plants.

Energy from plant biomass (bagasse and sugar cane straw in Brazil)

Brazil is currently the world's leading sugar cane producer. This is a milestone in the industrial development of the country, whose tropical areas have the highest ethanol production productivity on the planet. Despite this progress, Brazil still faces some challenges when it comes to the sectoral arrangements for utilising sugarcane bagasse and straw for energy production. A large-scale synergy between biomass plants and hydroelectric plants looks very promising if they can be included together in the energy reallocation mechanism, which ensures the optimisation of the interconnected hydroelectric generating park. The benefit of this synergy is materialised in a greater physical guarantee for the interconnected generator park. Studies show that almost half of Brazil's primary energy consumption today comes from sugar cane bagasse and straw biomass plants (Fiala & Bacenetti, 2012). There are environmental impacts resulting from the production and use of energy from biomass (Fiala & Bacenetti, 2012), but they are much smaller than the impacts caused by the use of fossil fuels.

There have been discussions in various international forums about the continued expansion of biofuel production in Brazil, leading to the loss of forests and biodiversity and ecosystems (Yan, n.d.).

Wind energy

Wind power is a renewable energy resource on a growing scale. Worldwide, the share of wind generation in the global electricity matrix has already reached 2.7 per cent in 2013. The United States has the highest proportion at 27 per cent, followed by China with 21 per cent. Brazil is the 15th country in wind power generation, corresponding to just 1% of the world's wind generation. The number of wind farms in Brazil has quadrupled in the last five years, from 70 in 2011 to 316 in 2015. This represents an expansion of 6,208 megawatts (MW) of installed capacity, which at the end of 2015 totalled 7,633 MW, having grown by an average of 109% a year since 2011 (MME, 2016). The north-east of Brazil accounts for around 60 per cent of all wind energy generated in the country, and the south accounts for 21 per cent of this total. It is estimated that wind power generation will continue to grow in Brazil, as more than 12 GW have already been contracted in the energy auctions organised by the federal government, and if this expansion continues, it could account for around 10% of all the electricity generated in the country by 2018.

Tidal energy

Tidal energy is not yet a significant part of the planet's energy matrix. It is basically harnessed in two ways: (i) by harnessing the potential energy from the difference in height between high and low tides; and (ii) by harnessing the kinetic energy of the waves. In France, the first tidal power plant was built in 1967 and connected to the national transmission network. The plant, with an installed capacity of 240 MW, harnesses the potential energy of the tides through a 750 metre long dam equipped with 24 turbines, built at the mouth of the River Rance in north-west France.

In Brazil, we have a wide range of tides in São Luís - Baía de São Marcos, in Maranhão - with 6.8 metres and in Tutóia with 5.6 metres, as well as in the estuaries of the Bacanga River (São Luís -MA - tides of up to 7 metres) and the island of Maracá (AP - tides of up to 11 metres). However, the topography of the coastline in these locations would require the construction of works of great environmental impact and high costs, making it unfeasible to harness tidal energy with the technologies currently available.

2.5 Primary energy sources and the electricity sector in Brazil

The broader debate on energy in Brazil has centred on the supply of electricity, ethanol and oil. But the Brazilian energy matrix is made up of different energy sources, with different environmental impacts, and a broader debate on the real benefits, productivity and environmental impacts associated with each energy source and its whole, could contribute significantly to greater awareness and better decisions on the expansion of energy supply and the Brazilian energy matrix. Around 39.4 per cent of the energy produced in Brazil comes from renewable sources and of the complementary 60.6 per cent, the largest share comes from oil and its derivatives, also accounting for 39.4 per cent. In other words, all the renewable energy consumed in Brazil is equivalent to the entire volume of oil also consumed. There are significant losses in the transformation of energy from its various forms to its final uses; these losses amount to around 10 per cent of all the energy produced in Brazil. This context justifies the implementation of a vigorous energy policy focused on expanding the use of renewable sources and promoting efficiency on the consumption side. Figure 1 illustrates the flow from energy sources to end use.

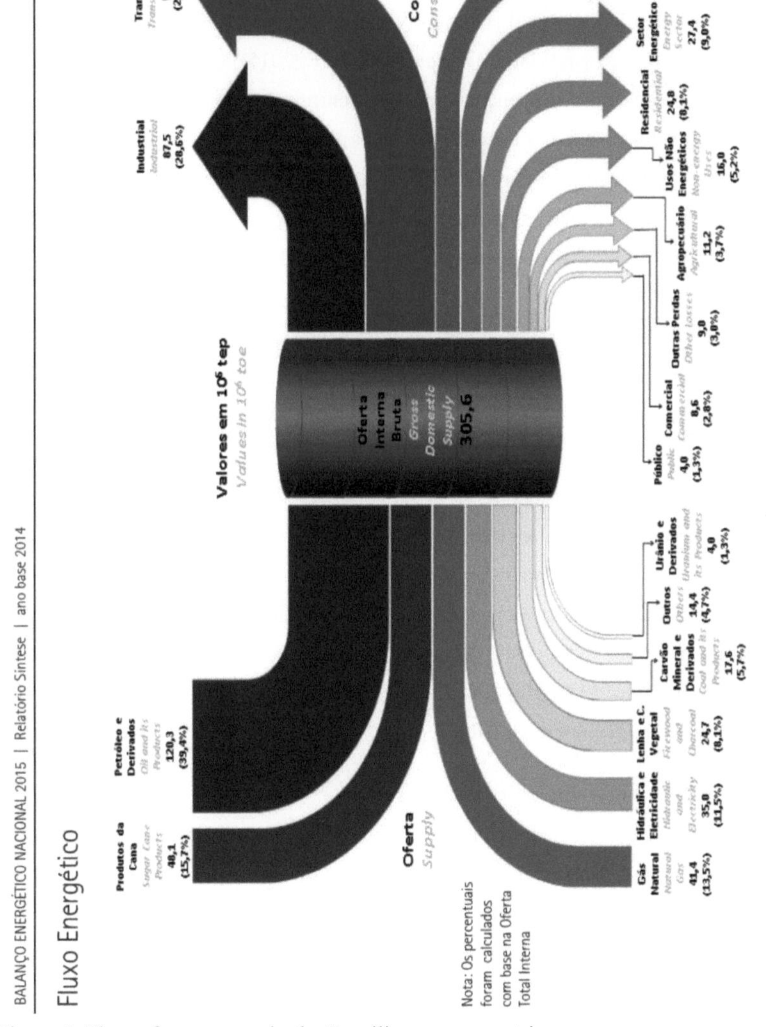

Figure 1. Flow of energy use in the Brazilian energy matrix

Around 65.2 per cent of all electricity consumed in Brazil is generated by hydroelectric plants, 13.0 per cent comes from natural gas, and only 7.4 per cent comes from biomass. Considering the volume of sugarcane bagasse and straw generated every year, which represents enormous potential for expanding electricity generation from this renewable source, it is also justified that energy policy in Brazil should incorporate strong incentives for the development of thermoelectric plants that use this waste from the sugar-alcohol industry, and that the optimisation of the use of the two primary sources, hydropower and biomass waste, could be based on an energy reallocation mechanism that compensates for the seasonality of the two sources.

Per capita electricity consumption in Brazil is currently in the order of 2,630 kWh/inhab/year, and has grown by around 2.5% (Electricity Statistical Yearbook, EPE, 2015). Historically, even during periods of economic recession, such as the one the country is currently experiencing, this *per capita* consumption has grown. This is yet another factor that justifies aligning policies to promote efficiency in energy consumption and incentives to increase the use of renewable energy sources that are abundant in the country, notably biomass, solar and wind, considering the commitments made by the Federal Government to the Framework Convention on Climate Change to limit greenhouse gas emissions in Brazil.

This context of the energy sector in Brazil, observed in the light of the international context of discussions on the energy issue, suggests that new generation projects and the promotion of energy efficiency developed by self-producers, independent producers, in short, by private initiative, can make a major contribution to the challenge of transitioning the energy matrix to a context of greater environmental and economic sustainability (Taylor et al., 2008).

Consideration of the problems related to energy consumption and projections such as energy efficiency, diversification of energy sources with the expansion of renewable energy sources, allowing greater participation in the decision-making process for the expansion of generation by private initiative and the end users themselves, the introduction of innovations, and visions that have been projected for the near future (Abeeku , 2007), will result in greater operational efficiency and sustainability, especially for companies in Brazil.

2.6The contribution of renewable energy sources

The use of renewable energy sources in Brazil has grown consistently. This can be seen in the latest editions of the National Energy Balance, although at a lower rate than possible. Studies on the evolution of the Brazilian energy matrix (Mill et al, 2010) drew attention to the success at that time of the programmes implemented in the mid-1980s in Brazil, the National Alcohol Programme and the Efficiency Programmes for the consumption of electricity and fossil fuels.

The focus of those policies at that time resulted in a consistent basis that could extend, if a new version of the energy policy is implemented, the benefits of expanding the use of renewable energy sources beyond the targets that have currently been proposed by the Brazilian government to increase the use of renewable energy sources by 15 per cent in the period 2006 - 2020. This target could be extended to 30 per cent by 2020 (Mill et al, 2010).

From a strategic point of view (Kankam and Shelter, 2009), they propose that building a more suitable energy future for Brazil's development requires a combination of policies that promote the vigorous utilisation of renewable energy sources. With the goal of advancing the use of renewable energy in mind, Brazil needs motivators to help decision-makers to invest in energy innovations overcome the obstacles to introducing innovations (Pearce, 1991). It is necessary to reduce taxation on imports and domestic taxation on equipment for utilising renewable energy sources. This development approach is in line with the recommendation of (Kankam and Aid, 2009) on the need for a combination of energy policy instruments for renewable energy sources to effectively promote development.

The advantages of using renewable energy sources cannot be minimised. Several advantages of renewable energy sources include job creation and ensuring that a nation is less vulnerable in its energy supply (Goldemberg, 1994). This can be easily understood in light of the fact that the capital invested in renewable energy sources is usually spent on materials and the people who build, operate and maintain the respective facilities, rather than on the purchase of energy.

CHAPTER 3

Method

3.1. Concept of environmental productivity of energy sources

The theories of production described by classical economic thinkers, including Adam Smith and Marx, considered labour power to be the main factor in generating economic value. When they were developed, those theories did not formally consider environmental resources as factors of production; however, they represent real limitations to economic growth, limitations that vary according to the technologies and energy policies adopted.

Commercial energy flows and the environmental resources used in their production are real factors of economic production, and both limit economic growth, just as the availability and timely use of capital in infrastructure investment also limits economic growth.

The "environmental productivity" of energy sources is an element that defines a metric for evaluating energy sources and matrices in terms of their environmental efficiency, this efficiency being understood as the ratio between the flow of energy offered for consumption and the amount, or extent, of the biosphere and atmosphere needed to supply the corresponding energy.

Measurements of the demand imposed by man on ecosystems have been made, generally taking measurements of the areas needed for energy production facilities as an expression of the size of the "ecological footprint" left by man in satisfying his needs, however, there is also criticism of the calculation made by taking only the area as the explanatory variable of the impact. Taking into account the energy flow of each energy source that makes up the energy matrix, and all the physical extensions of the biosphere and atmosphere necessary for the commercial production of the respective energy flow, is a more consistent contribution to both academic discussion and energy policy governance.

The study's approach considers "environmental productivity" as the ratio between the Energy supplied by a given energy source or energy matrix and the sum of the Areas impacted by the respective energy production, comprising: (i) the Area used by the respective energy production facilities, qualified according to their importance and priorities established for biodiversity conservation in the respective region, and (ii) the equivalent Area required to form and maintain carbon stocks that correspond to the volume of greenhouse gas emissions resulting from energy production with the respective source. These "carbon stocks" could be natural, although subject to human intervention for their formation and expansion, such as native forests, whose expansion and regeneration could represent additional stocks that compensate for emissions, or they could be artificial, using dedicated CO_2 capture and storage technologies, next to emitting sources such as thermoelectric power stations, in whose arrangements the carbon captured and stored in underground reservoirs, or mixed with seawater, using technologies still under commercial development, such as: (i) Post-combustion technology, in which CO_2 is separated from the gases emitted by injecting them into a liquid column containing solvents such as Ammonia, resulting in the separation of CO_2 that can be stored in underground reservoirs, or mixed with seawater; (ii) Oxyfuel technology, in which the air-fuel mixture is enriched with pure oxygen, resulting in emissions with a higher content of CO_2 and water vapour, in turn making it possible to pump the CO_2 into a reservoir; and, (ii) Pre-combustion technology, mainly applicable in coal-fired power stations, in which coal gasification is employed, producing a synthetic gas rich in carbon monoxide and hydrogen, in turn destined for reaction with water, forming CO_2 which is then captured and stored, plus hydrogen, and the latter burnt in gas turbines to produce electricity. In this study, we used the natural reservoirs of native forests as a reference for calculating the environmental productivity of energy sources.

The environmental productivity of an energy source is an indicator of the environmental effort required from the biosphere and the atmosphere to generate energy through the respective source. The "environmental productivity" of a given energy source z is expressed as the ratio between the amount of energy generated by the source in a given time interval and the extent of the impacted areas in the biosphere and atmosphere, the latter expressed in equivalent area in the biosphere, as illustrated in Equation (1).

$$P_i = \frac{EU_i}{B_i + A_i} \qquad (1)$$

Where: P_i = Environmental productivity of source i

EU_i = Energy generated by source i

B_i = Area of the biosphere impacted by the installation of source i

A_i = Equivalent area of native vegetation cover needed to compensate, in stock, for the carbon emitted in the form of Greenhouse Gases by source i, in a given period of time, needed to generate £77, of energy

In an energy matrix made up of *n* generating plants, their environmental productivity can be calculated using expression (2).

$$P = \frac{\sum_{i=1}^{n} EU_i}{\sum_{i=1}^{n} (B_i + A_i)} \qquad (2)$$

3.2. Equations for applying the concept of environmental productivity of energy sources

Equation (1) can be broken down into two factors: the average capacity factor of the installation of energy source i, and the environmental effort required to install it.

Equation (3) illustrates this breakdown of factors.

$$P_i(t) = \frac{EU_i(t)}{CI_i} \times \frac{CI_i}{B_i + A_i} \qquad (3)$$

Where:

Cl_i = Installed capacity of energy source z

The other factors in equation (3) were defined in equation (1).
Considering time intervals of 1 year for the energy data, $EU/Cl_{i\,i}$ is in fact the capacity factor of the energy source on an annual basis. Equation (4) is the form adopted in the calculation.

$$P_i = FC_i \times \frac{CI_i}{B_i + A_i^T} \qquad (4)$$

Where:

P_i = Environmental productivity of energy source *i*

FC_i = Capacity factor on an annual basis for energy source *i*

B_i = Area of the biosphere impacted by the installation of source *i*

$\acute{A}_i^T$ = Equivalent area of the biosphere needed to remove from the atmosphere and keep in stock an equivalent volume of carbon emitted in the form of greenhouse gases during plant operation.
Therefore, the environmental productivity of energy source *i* is equal to the product of its average capacity factor, the ratio between its installed capacity and the extent of the impacted areas, the biosphere and the equivalent area of the atmosphere, during a given period of time *T*.

3.3 Environmental productivity and ecological footprint

Adopting greater "environmental productivity" is equivalent to adopting a smaller "ecological footprint", but the methods for determining one and the other are different. Determining the "environmental productivity" of energy sources uses the design parameters of energy generating plants, and is more directly associated with information commonly used by decision-makers in investment projects for new electricity generating plants and energy policy makers; while determining the "ecological footprint" uses the "ecological footprint".

The "ecological productivity" of energy-generating plants or energy matrices requires their contextualisation in relation to the planet's capacity for regeneration, using parameters external to the decision-making processes of energy-generating plants, requiring the formal incorporation of these external parameters into the respective decision-making processes. The "environmental productivity" of energy sources can be incorporated into the decision-making processes for the implementation of new energy sources, as a limiter for the approval of projects, and the respective acts of authorisation by the authority responsible for authorising or granting the new plant, as a floor, or minimum value of "environmental productivity", to be observed in the project approval processes, and respective acts of concession or authorisation.

"Environmental productivity", of any source whose production is sustained from natural resources, considers absolute quantities of extensions of the biosphere and atmosphere to sustain production; while the "ecological footprint" considers relative capacities of the biosphere associated with production, compared to the global capacity of the planet to sustain said production.

The "ecological footprint" means the ratio between the quantity demanded of a product associated with the biocapacity of the planet required to sustain said production, and the available biocapacity (Borucke et al, 2013).

In elementary form, the "ecological footprint" is expressed by equation (5).

$$EF = \frac{D}{Y} \qquad (5)$$

where: *EF* corresponds to the "ecological footprint"

Give the annual demand for a product

Y is the global annual production of the same product

The "ecological footprint" as conceptualised does not consider non-renewable resources in its calculation, as it only considers renewable resources in the biosphere.

The "ecological footprint" assumes that human occupation and infrastructure often occupy arable and fertile areas. It considers that the installation of infrastructure implies giving up agricultural production. Thus, areas occupied by energy generation plant infrastructures reduce the regeneration capacity of the territory in which they are located, while at the same time reducing the agricultural production.

productivity factor of arable areas.
The "ecological footprint" of these infrastructures, particularly for hydroelectric plants, is calculated using equation (5) (Wackernagel et al, 2005).

$$Footprint_{buiit\text{-}up}\,(gha) = Area_{buiit\text{-}up}\,(ha) * Equivalence\ Factor_{buiit\text{-}up}\,(gha/ha) * Yield\ Factor_{cropiand} \quad (5)$$

Where:

$Footprint_{buiit\text{-}up}$ (gha) =Ecological footprint in global hectares

$Area_{buiit\text{-}up}$ (ha) =Area occupied by the power generation plant in hectares

Equivalence $Factor_{buiit\text{-}up}$ (gha/ha) =Equivalence factor in global hectares per hectare

Yield Factor cropiand =Area productivity factor

The areas used to form the hydraulic reservoirs of hydroelectric power plants can vary greatly in their productivity, and are generally not well documented. Due to this information gap, the aforementioned authors adopted a global average equivalence factor equal to 1.0 and a productivity factor also equal to 1.0. And considering that electricity consumption is recorded and better documented than land use in reservoir areas before the construction of hydroelectric plants, they use a constant factor to convert energy use into area, according to equation (6):

$$Footprint_{hydro\ area}\,(gha) = Energy\,(GJ) / constant\,(GJ/ha) * Equivalence\ Factor_{hydro\ area}\,(gha/ha) \quad (6)$$

Where:

$Footprint_{hydro\ area}$ (gha) =Ecological footprint

Energy (GJ) =Energy generated in Gigajoules per hectare

constant (GJ/ha) =Energy use conversion constant in area

Equivalence Factorhydro $_{area}$ (gha/ha) = Equivalence Factor

The constant factor of conversion of energy use into area was taken as an average of the twenty largest hydroelectric plants on a global basis (WWF, 2000). And empirically the "ecological footprint" was reduced tenfold for the cases of installed hydroelectric power plants

in mountainous areas, such as Switzerland and Norway, to compensate for the fact that hydroelectric plants in these areas need smaller areas for their reservoirs, and the fact that mountainous areas are generally not as arable as river valley areas.

Considering this example of how to calculate the "ecological footprint" for infrastructures, which can be extended in general to power generation plants from the various types of sources, with the exception of nuclear power, due to the diverse and significantly different nature of the environmental impacts and risks associated with it; and comparing it with the concept and equations for calculating the "environmental productivity" of energy sources, we believe that both concepts are metrics that can be used in efforts to reverse the trend of increasing environmental impacts associated with electricity generation.

The "environmental productivity" of energy sources can, however, be useful to help in these endeavours, based on references limited to the decision-making processes for installing new energy generating plants. And it can also help in making specific decisions to decommission power generation plants that should be eliminated due to their very high environmental impacts, evidenced by their very low "environmental productivity", as is the case with some coal-fired plants and some plants that consume liquid fossil fuels.

CHAPTER 4

Application of the concept of environmental productivity to the sources of the National Interconnected Power System (SIN) and aggregates of the Brazilian energy matrix

The calculation of the environmental productivity of energy sources can then be based on the source's capacity factor, its installed capacity, and the total area impacted by its installation and operation, which comprises the area of the biosphere impacted by the installation of the source, and the equivalent area of native vegetation cover (of the biosphere) corresponding to the carbon stock that must be added each year to compensate for the impacts on the atmosphere resulting from the emissions of gases that cause the intensification of the greenhouse effect resulting from the operation of the energy source.

The areas used to install the generating plants for energy sources, as well as the installed capacities, are information that is generally used in environmental authorisation and licensing processes.

This session presents the application of the concept of "environmental productivity of energy sources" and its application to the case of the SIN's electricity generation plants, and the aggregates of the main energy sources used in Brazil, namely sugar cane energy, hydroelectricity and oil.

4.1. Environmental productivity of SIN energy sources

Compatibility between information on areas, installed capacities and energy is a challenge for the consistent calculation of "environmental productivity", since more precise records of affected areas are generally obtained from environmental impact studies associated with the environmental licence for the respective installation. Environmental studies have criteria that may vary in some respects from one study to another, although they are all based on the same legal precepts; this was the biggest challenge faced in obtaining a reliable and comparable database for the plants in the National Interconnected Electricity System (SIN) and for the aggregates of the energy matrix, in the calculation tests developed.

The Ministry of the Environment, in fulfilment of Brazil's commitment to the Convention on Biological Diversity signed in 1992, has carried out a wide-ranging assessment of ecosystems in Brazilian territory, establishing priority areas for biodiversity conservation, resulting in a classification that makes it possible to differentiate the importance and equivalence between classified areas. These priority areas for biodiversity conservation were originally mapped and geo-referenced, based on the provisions of MMA Ordinance No. 126 of 27 May 2004, and periodically reviewed under the terms of that Ordinance, with a first review having been carried out based on the provisions of MMA Ordinance No. 09 of 23 January 2007.

The mapping resulted in the classification of areas of biological importance represented on maps published by the MMA, with relative importance graded by multiplier factor, with the notation shown in Table 2. The relative ranking of high importance is presented in three different colours, to distinguish between biomes and micrographic basins which, although they have the same relative ranking, have different characteristics.

Table 2. Classification of priority areas for biodiversity conservation

Colour associated with priority	Importance	Relative graduation
Red	Extremely high	10,000 x Affected area
Orange	Very high	8,000 x Affected area
Yellow	High	5,000 x Affected area
Grey	High	5,000 x Affected area
Purple	High	5,000 x Affected area
Blue	Insufficiently known	1,000 x Affected area

Source: Ministry of the Environment, 2004

The energy sources of the SIN were represented on a map of Brazil according to their geographical location, and the superimposition of this map with the map of priority areas resulted in the map of the location of energy sources according to the importance of their location for the preservation of biodiversity. The maps were superimposed using georeferenced image files and, using the Google Earth application, the resulting maps are shown in Figures 2, 3, 4 and 5.

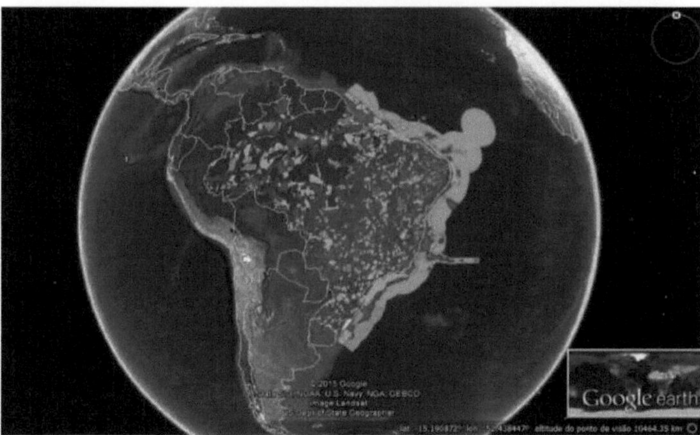

Figure 2 - Map of priority areas according to their importance for biodiversity

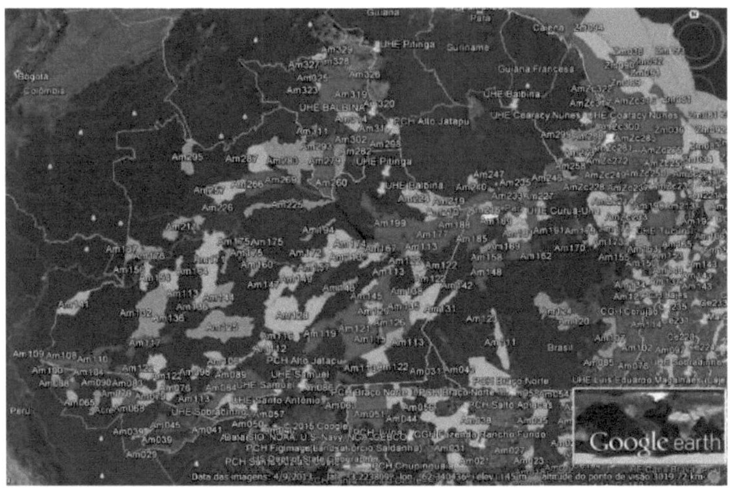

Figure 3 - Map of energy sources and priority areas for biodiversity conservation in the northern region of Brazil

Figura 4. Map of energy sources and priority areas for biodiversity conservation in northeast Brazil

Figura 5. Map of energy sources and priority areas for biodiversity conservation in the southeast and south of Brazil

Based on the locations of the electricity generation sources and their superimposition on the map of priority areas for biodiversity conservation, the area considered in the calculation of environmental productivity was adjusted, using multiplicative factors equal to those shown in Table 2. It should be noted that this consideration is valid for the period of time covered by the life cycle of each plant, and therefore the calculation of the environmental productivity of the energy sources must also refer to the period covered by the respective life cycle.

The areas used to install hydroelectric plants were obtained from the database of generation facilities kept by ANEEL (2015).

For thermoelectric plants, whose installation areas are not fully included in the aforementioned database, and considering that the respective information is not available from reliable sources, the representative factor of 4.54 x 10' ha/MW was used, corresponding to the average of specific values found in the technical literature on thermoelectric plants.

The environmental impacts of wind farms are generally localised, with the most important interference related to birdlife, noise and electromagnetic interference. The impact on the biosphere and atmosphere is considered to be proportional to the size of the area used to install the wind farm. This area depends on the location, particularly the topography of the area, obstacles and particularities of the relief; and is generally considered to be equivalent to 4 to 8 times the rotor diameter for each wind turbine. In Brazil, the parameter of 6 MW/km^2 (CEPEL, SÁ and DUTRA, 2011) has been adopted, which is equivalent to 1.67x10$^{'2}$ ha/kW.

In the case of solar plants, the environmental impacts are also localised, with the most important interference related to the vegetation cover lost in the installation area. The strain on the biosphere is then considered to be proportional to the extent of the area used to install the solar plant. The different solar plant technologies, including thermosolar and photovoltaic, require different areas per unit of installed capacity. Photovoltaic technologies in particular have shown a progressive increase in power density per unit of installation area. The parameter of 0.65 ha/MW, equivalent to $6.5x10^{-4}$ ha/kW, is estimated to be representative for solar plants at the current stage of technology, based on projects yet to be implemented in Brazil.

The areas corresponding to carbon stocks, equivalent to greenhouse gas emissions in the operation of the energy sources over the concession or authorisation period considered (30 years), were calculated based on characteristic carbon stock values found in tropical forests in the southeast of Brazil, which represent a mix of tree species from the Atlantic Forest and Cerrado. An average factor of 373.31 tCCEeq/ha was verified in a forest restoration project carried out in the state of São Paulo in 2007, for the restoration of Permanent Preservation Areas (APP) around hydroelectric power stations on the Tietê River and around the Água Vermelha hydroelectric power station on the Grande River (AES Tietê, 2007). This factor represents the amount of carbon removed from the atmosphere and stored in the trunks, roots, branches and leaves of trees. There is another significant amount of carbon that is incorporated into the organic matter on and in the soil, which can raise the factor to around 600 tCO2eq/ha (AES Tietê, 2007). The same factor was adopted as representative of areas in the north, northeast and south regions, although it is known that there are differences between the representativities. However, in order to adopt different factors for each region, it would be necessary to make adjustments to the values found in the literature available in Brazil, in order to establish a reliable and comparable basis, work that could not be carried out in the time used to develop this Thesis. Notwithstanding the purpose established here, this effort is recommended for future developments.

Greenhouse gas emissions from the operation of SIN plants occur essentially in atmospheric emissions from thermal power plants, and methane (CH4) from hydroelectric plant reservoirs. Emissions associated with the manufacture of equipment used in the energy sources, as well as their construction works, were not considered in this approach, and this should also be noted as an aspect to be developed in the refinement of the environmental productivity calculation. The other sources, wind and solar, are only associated with atmospheric emissions in the manufacture of equipment and installation work, so they were not considered in the calculation of emissions from their operation.

For emissions from thermoelectric plants, average emission factors representative of all thermoelectric sources connected to the SIN were adopted, based on data published by (EPE, 2015). These factors are shown in Table 3.

Table 3. Average emission factors of thermoelectric plants in the SIN

Fuel	tC/TJ	tCO /MWh$_2$
Natural Gas	15,30	0,70

Coal	25,80	1,38
Fuel Oil	-	-

Source: Empresa de Planejamento Energético, 2015

For methane emissions in the reservoirs of hydroelectric power plants, factors obtained from measurements made in the hydraulic reservoirs of hydroelectric power plants in the north, southeast and south (PINGUELLI et ah, 2008) were adopted, whose measured values varied widely, and the use of these factors should therefore be cautioned due to their short history and the variations found. However, the work consulted reveals trends and behaviours that explain, albeit preliminarily and with the caveats pointed out, the reasonableness of adopting factors that are representative of CH4 emissions over periods of time from the formation of the reservoirs. We have always considered the first concession period for hydroelectric plants, as this is the period in which the most intense emissions occur due to the decomposition of submerged organic matter with the formation of the reservoir, adopting the factors shown in Table 4.

Table 4: Estimated emission factors from hydroelectric plant reservoirs

Period	Kg C/km /day^2	Kg CO - /ha/day$_{2eq}$	KgCO . /ha$_{2eq}$
1st and 2nd year°	50	10,5	7.665,0
3° to 10th grade	20	4,2	12.265,0
11th to 30th years	5	1,1	8.030,0
Total			27.960,0

Source: Oecologia Brasiliensis, 12(1): 116-129,2008

The environmental productivity values of the SIN's energy sources were calculated using a spreadsheet made with Excel software, a file of which is available here.

together with the electronic archive of this document. Table 5 shows an extract with representative data from historical periods during which relatively homogeneous series have been verified. This extract shows the significant increase in environmental productivity with the introduction of wind sources in the first decade of the 21st century, and in the first half of the second decade.

Table 5. Environmental productivity of SIN sources in historical evolution and by type of source

Plant	Year	Primary energy source	Granted power (kW)	Physical guarantee (averag	FC (%)	Impacted area of the Biosphere (ha)	Atmosphere impacted area (ha)	Total area impacted	Environmental productivity (kW/ha)

				e MW)					
Canastra	1905	HPP	44.800	24,00	53,57	358.982,44	16.728,58	375.711,03	0,0639
Tear	1925	HPP	22.000	11,84	53,82	176.286,02	8.214,93	184.500,95	0,0642
Fonte Nova	1940	HPP	131.988	104,00	78,80	1.057.619,98	49.285,09	1.106.905,07	0,0940
Golden Waterfall	1959	HPP	658.000	415,00	63,07	5.272.554,66	245.701,05	5.518.255,71	0,0752
Spring Harbour	1999	HPP	1.540.000	1.017,00	66,04	12.340.021,54	575.045,00	25312057,87	0,0787
Uruguaiana	2000	UTE	639.900	565,10	88,31	29,05	1.576.472,51	1.576.501,56	0,3585
		(Natural Gas)							
Osório Wind Farm	2006	EOL	50.000	17,71	35,42	835	0	835	21,2096
Back from the river	2010	EOL	42.000	19,84	47,24	701.4	0	701,4	28,2863
Dracena 1	-	UFV	30.000	5,90	19,67	19,50	0	19,50	302,5641
Solar Caetité	-	UFV	29.970	6,60	22,02	19,48	0	19,48	338,8003

Note: 1. Table 5 shows an extract of some of the SIN's energy sources, according to their commissioning date; it was extracted from an extensive spreadsheet containing a complete list of the SIN's sources, with a very high number of lines.

2. The electronic spreadsheet containing the complete list of SIN sources is available together with the document in this Thesis.

Graph 1 illustrates the evolution over time of the Environmental Productivity of the SIN's sources, showing a marked increase in environmental productivity in the first decade of the 21st century, when wind farms began to be installed in Brazil.

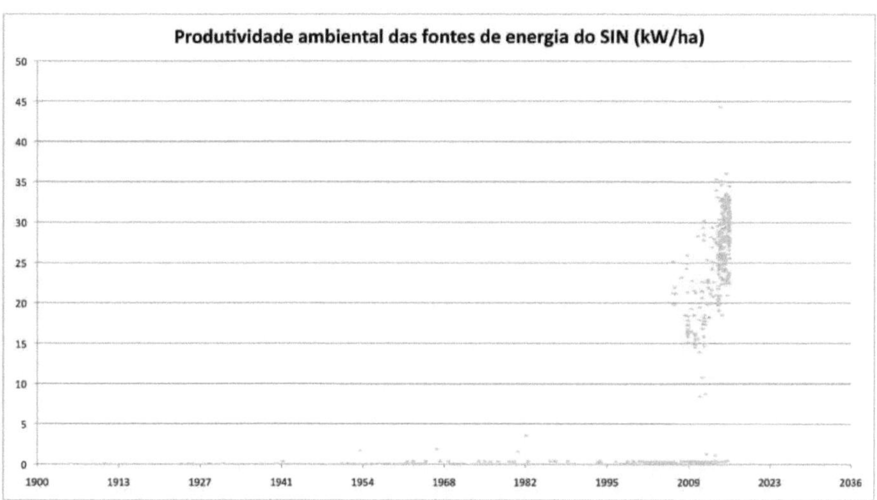

Graph 1. Environmental productivity of SIN sources in historical perspective

Graph 2 illustrates the dispersion of environmental productivity values of SIN sources in the first two decades of the 21st century, indicating a significant increase in environmental productivity values in the second decade.

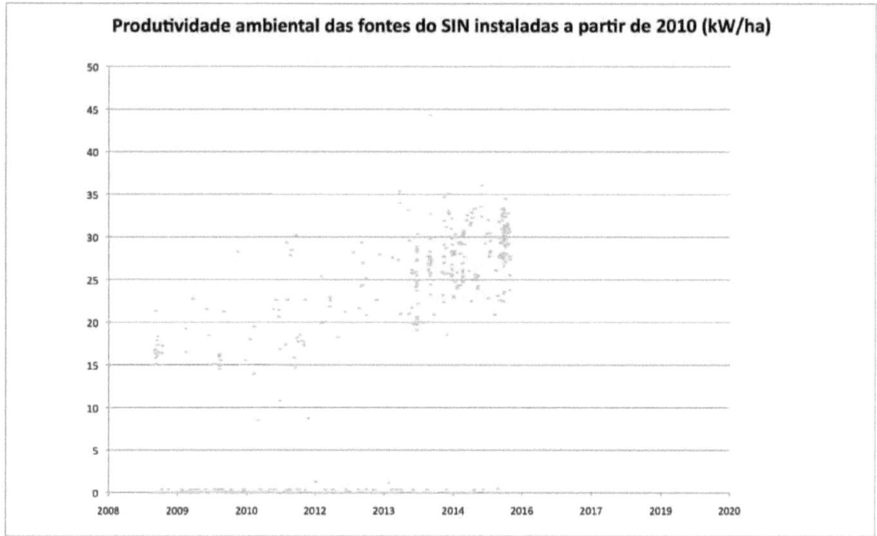

Graph 2. Environmental productivity of SIN sources in the first two decades of the 21st century

4.2 Environmental productivity of the main aggregates in the Brazilian energy matrix

The concept of environmental productivity of energy sources can be applied to aggregates of the energy matrix, although in this case it was not possible to incorporate the differentiation of priority areas for biodiversity conservation into the calculation, nor the emissions from hydroelectric plant reservoirs, due to the unavailability of the respective data at the aggregate level of the energy matrix. In addition, energy production and supply systems have portions of energy losses inherent in their production and transport technologies, until they are made available for final use, making it necessary to express them in a common base of energy made available by source, so that the comparison between them is not distorted.

The concept of the environmental productivity of energy sources does not imply that its calculation takes into account useful energy, which incorporates the efficiencies of the end uses of the various forms of energy. It must be calculated on the basis of the energy actually made available for end uses. The most consistent database of energy aggregates by source, which prominently displays energy losses (before final consumption) is the Useful Energy Balance (BEU), which was last published by the MME in 2006; however, the BEU also incorporates the energy yields of the final uses of the forms of energy, whose evolution has shown a similar profile to the evolution of the efficiencies of the final uses of the various forms of energy.

It is therefore considered that the trends in the environmental productivity of primary energy sources, calculated on the basis of their aggregates, can be reasonably observed from the calculation made on the basis of the useful energy aggregates presented in the BEU.

To compare energy sources, the unit of tonnes of oil equivalent (toe) was adopted, resulting in the unit of tonnes of oil equivalent per hectare (toe/ha) for the environmental productivity of aggregate energy sources.

It is therefore reasonable to visualise trends in the environmental productivity of the aggregate sources studied, in a way that is consistent with what would be expected from a detailed study of individual sources.

Three of the main commercial sources of energy in Brazil are: (i) sugar cane, (ii) hydroelectricity, and (iii) oil, of which only the portion used to produce final energy, whatever its use, is considered. Together these sources account for around 70 per cent of the domestic energy supply.

The calculations of the environmental productivity of the aggregates of sugar cane, hydroelectricity and oil sources, as well as a large part of the electricity generation plants that make up the SIN's generating park, are presented below.

Table 6 shows the Primary Energy values obtained from the BEU, presented by the MME together with the National Energy Balance for 2006, Base Year 2005, the last year of the historical series of aggregates published from data collected from all primary energy sources, and consisting of aggregates and useful energy. In the years since 2006, the MME has published the National Energy Balance, containing a brief commentary on the useful energy balance, but without adding updated data.

Table 6 - Primary Energy

Primary Energy	(Thousand toe)		
	1990	1994	2004
Sugarcane	1845	22010	2938
1			5
Hydroelectricity	17770	20864	2758
			9
Oil	32550	34446	7664
			1

Source: MME, 2006

Table 7 shows the energy yields, also obtained from the BEU published in 2006, for the aggregates of the energy sources considered. The values of the yields of the energy sources at an aggregate level are presented here as an adjunct to the characterisation of the aggregate sources.

Table 7 - Energy Yields

Energy yields	(%)		
	1990	1994	2004
Sugarcane	65	71,6	76,7
Hydroelectricity	58,1	64,3	68,8
Oil	35,6	40,5	43,4

Source: MME, 2006

Table 8 shows the calculated Energy Generated values for the aggregate sources.

Table 8 - Energy generated

Generated Energy	(Thousand toe)		
	1990	1994	2004
Sugarcane	11993	15759	22538
Hydroelectricity	10324	13416	18981
Oil	11588	13950	33262

Source: MME, 2006

Table 9 shows the areas used to install the generating capacities of the energy source aggregates and, in the case of oil consumed for energy purposes, the areas indicated were calculated based on the correlation between the volume of CO2 emissions and the area of native forests that would have to be planted to remove emissions from the atmosphere and store the respective volume of CO2 emissions.

The correlation between native forest areas and the volume of emissions was considered based on values characteristic of tropical forests in the southeast of Brazil, representing a mix of Atlantic Forest and Cerrado forests. This correlation is represented by the average factor of 600 tonnes of CO2 equivalent per hectare, including the carbon stock in the soil.

Table 9 - Areas used by aggregates from aggregate sources

(10^6 ha)

	1990	1994	2004
Sugarcane	4,29	4,36	5,57
Hydroelectricity	2,53	2,78	3,58
Oil	33,89	39,42	NA

Sources: MME, 2006; and AES Tietê, 2007

Calculating the environmental productivity of the three sources considered, aggregated in the Brazilian energy matrix, results in the values shown in Table 10.

Table 10 - Environmental productivity by energy source aggregated in the matrix

Environmental productivity by source	(toe/ha)		
	1990	1994	2004
Sugarcane	2,795571	3,61445	4,046320
Hydroelectricity	4,080632	4,825899	5,301955
Oil	0,341930	0,353881	NA

The research carried out sought to test the concept of "environmental productivity of energy sources" in the SIN, considering the availability of more complete and consistent data on electricity generation facilities; and to test the application of this concept in the aggregate of three of the main energy sources in the Brazilian energy matrix. Determining the boundary conditions of an energy matrix based on energy and environmental logic for Brazil, as well as conclusions and recommendations on the use of the concept tested as a subsidiary element of an energy policy based on the environmental productivity of energy sources, complete the objective set for this Thesis.

Extending this study to other national energy matrices, and evaluating them as a whole, would make it possible to better visualise the potential contribution of a global energy policy to the effort to reduce emissions, and would make it possible to evaluate the economic and environmental costs and benefits of developing a virtuous cycle of transition in the global energy matrix, based on investments in new renewable energy generation plants, made possible by regulatory incentives, tax reductions or exemptions and, above all, financing, which could be guaranteed by a global energy policy, using implementation mechanisms that are also global, committed to within the framework of the Framework Convention on Climate Change.

CHAPTER 5

Discussion in the light of the concept and calculation results

The energy sector in Brazil, although considered to be one of the lowest emitters of greenhouse gases (GHG), nevertheless presents an incongruity from an energy and economic point of view, with growing vulnerabilities resulting from uncontrollable factors such as the accentuation of seasonalities currently seen in hydraulic inflows to the interconnected hydroelectric system.

This increase is probably due to climate change as a result of global warming (Metz, 2002). This inconsistency could be minimised, in order to preserve energy security, by aligning energy and environmental policies, aimed at adjusting the energy matrix based on establishing minimum environmental productivity for new energy sources that may be licensed.

From a global perspective, the energy sector accounts for more than 60 per cent of total GHG emissions, which increase the greenhouse effect by applying technologies that have already been used throughout the history of industrial development in developed countries, and which are still currently being used in the recent development of emerging economies (Preto & Weisel, 2010).

In developed countries, the energy sector includes the generation of electricity from oil derivatives, coke ovens and they account for more than 60 per cent of gas emissions, which in turn lead to an increase in the greenhouse effect (Gerber & Opio, n.d). On the other hand, there are a few countries whose share of emissions from the energy sector is relatively low, such as Brazil and Argentina, whose percentages are below 30 per cent (Preto & Weisel, 2010). This share is subject to substantial variation in the comparative context of various countries, as shown in Graph 3.

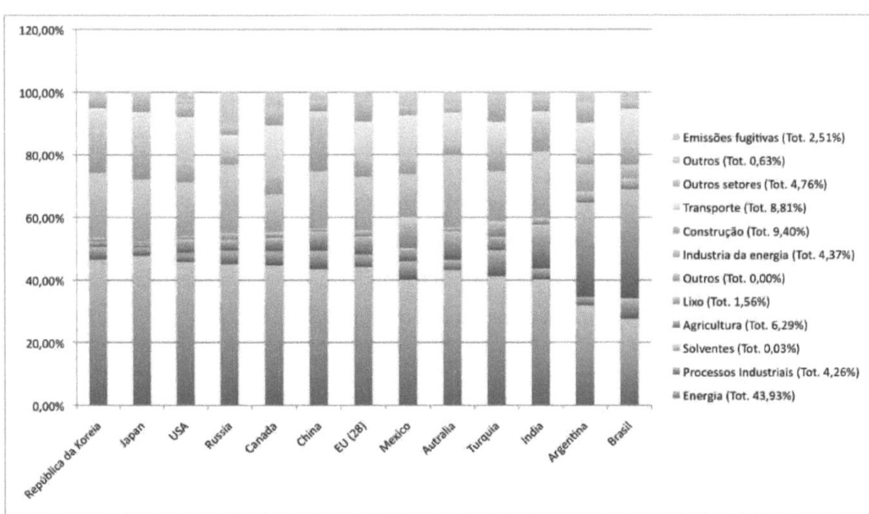

Graph 3. Participation of economic sectors in country emissions

Source: UNEP 2012. The Emissions Gap Report 2012. United Nations Environment Programme (UNEP), Nairobi (Gerber & Opio, n.d.)

In general, developed countries have been recognised for their efforts to replace conventional sources of energy with other sources with lower carbon emissions. In addition, increasing efficiency in energy consumption is also an important alternative for a consistent reduction in greenhouse gas emissions. Furthermore, developing countries whose greenhouse gas emissions have been predominantly associated with changes in land use, mainly those uses related to agriculture and deforestation, in order to maintain a myriad of activities (Gerber & Opio, n.d), have substantial potential for reducing energy consumption by increasing efficiency. There are alternatives for increasing energy supply, such as adopting policies that promote a greater supply of energy from renewable sources with low environmental impact, which can provide a more accelerated transition for their energy systems, with no re-editing of technological routes that prevailed in the development of more developed economies (Metz, 2002).

In this context, fundamental issues have been debated both locally and globally:

1) What would be the desired energy system for the planet as a whole and for each specific country in a context of increasing environmental restrictions, and what should a global energy policy look like to enable the transition of the global energy matrix towards lower environmental impacts?
2. What common global effort could be successful in terms of reducing global emissions in order to preserve the temperature increase on the Earth's surface at 2° C by 2020, in order to fulfil

the targets agreed for the Conference of the Parties to the Convention on global climate change?

3. Since 2010, the United Nations Energy Programme (UNEP) has indicated trends to reduce global emissions, such as the commitments made by the parties at the COP, and has pointed out a large gap for 2020 in greenhouse gas emission reductions in reference to what is needed to reach the target of 2° C. But there are also technical alternatives to eliminating this gap if swift action is taken, principally the adoption of energy policies to promote greater use of renewable energy sources in place of other intensive sources of greenhouse gas emissions (Goldemberg, 1994). These policies should be adopted through a global agreement with an appropriate instrument for the transfer of technologies, including an effective instrument for financing their implementation, which could stand to make significant contributions to a global effort to reduce greenhouse gas emissions in the short term. What instruments could be used to implement a global energy policy that could be the subject of an agreement between nations at the COP within the scope of the Convention on Global Climate Change?

A combination of energy, economic and environmental issues has characterised the problems in energy sectors around the world; solutions have been proposed with greater or lesser emphasis on each of these issues (Smithson, 1989).

However, with a detailed analysis of the case of Brazil and from this analysis examining the context of emerging countries in general, and also analysing the situation of energy sectors in developed countries, it can generally be concluded that energy policies in reality are predominantly oriented on the basis of a combination of economic, energy security and trade policy issues rather than being oriented towards energy systems with a lower environmental impact (Smithson, 1989).

Nevertheless, there is already considerable technological development that allows for more ambitious energy policies, such as those that accelerate the decarbonisation of the energy sectors on a global scale (Lough, 1996). The installation of new, less emissions-intensive energy generation plants, made possible by these policies, would be able to create a new virtuous cycle for new investments, something very desirable in the current context of depression in several economies (Smithson, 1989).

The International Energy Agency continually assesses energy-related technological advances. Energy technology outlooks were published in 2012, including very positive developments in renewable source technologies. Of particular note is a significant reduction in the cost of photovoltaic energy generation, as commented on and illustrated in Annex III.

In Brazil, the expansion of energy generation takes place through annual auctions, when new energy sources are contracted at the lowest price offered by the energy producer at the Auction, and are contracted by all the energy distributors responsible for supplying electricity to captive users. A sharp reduction of more than 20 per cent was seen in prices for wind energy negotiated at auctions. Wind energy represents a new source and new technology in Brazil. The development of wind turbine technology using towers over 80 metres and, more recently, over 100 metres, and the development of turbines that make more effective use of winds between 4 and 7 metres per second, has led to a significant reduction in the prices negotiated at auctions. Table 11 illustrates the evolution of the price of energy negotiated in Brazil.

Table 11. Electricity prices negotiated at auctions in Brazil by source (R$/MWh), at December/2015 prices.

Year	Hydroelectric	Biomass	Wind power
2007	239	245	265
2008	239	250	218
2010	197	203	184
2011	123	140	142
2012	120	-	118
2013	137	162	143
2014	138	233	143
2015	168	298	206

Source: Chamber of Electric Energy Commerce (CCEE), 2015

Solar and wind energy sources have significant potential for expansion not only in Brazil's energy system, but also in the global energy system, with effective reductions in greenhouse gas emissions. In addition, evidence of cost reductions indicates the possibility of adopting a global energy policy to grant credit for the replacement of sources that are intensive in greenhouse gas emissions through criteria that ensure an effective reduction in such emissions, supported by differentiated financing and taxation, which in turn makes them economically viable when compared to conventional solutions (Preto & Weisel, 2010).

Other relationships of different natures have also contributed to the price performance of solar and wind energy: large investments in photovoltaic module production facilities in China, consequently resulting in a huge increase in production, an increasing number of photovoltaic systems installed in Germany, resulting in a large reduction in installation costs, and better utilisation of idle reserve capacity for wind turbine manufacturing in Europe, as a result of the 2008 financial crisis, which allowed demand for wind turbines to be met at reduced prices in Brazil (Smil, 2008).

However, considering the reference marginal costs of expanding energy supply in Brazil and Europe, respectively 88 USD/MWh[1] and 52 USD/MWh, reduced prices for solar and wind power generation, as well as considering the respective intensities of greenhouse gas emissions in energy systems, the

[1] The marginal cost reference for the expansion of energy supply in Brazil is determined on the basis of the ten-year energy plan drawn up by the Federal Government, through the Energy Research Company (EPE).

costs corresponding to emission reductions by replacing conventional sources with zero emission sources are estimated at 115 USD/kg CO2eq in Brazil and 117 USD/kg CO2eq in Europe.

These costs are quite low for reducing emissions when compared to the costs incurred by the global economy for mitigating the impacts caused by climate change, which would be more intense in the case of the scenario of an insufficient reduction in emissions as established by the Global Convention on Climate Change, with the goal of containing the increase in the Earth's surface temperature to 2°C by 2020 (Dinda, n.d).

The expansion of the energy supply through these and other lower-emission sources, such as biomass and small hydroelectric power stations, competes with conventional sources for investment, using lower cost criteria. However, these criteria do not take into account the real cost of the sources, including the socio-environmental costs of their use. The decision-making processes for such investments are basically based on the principles of lower cost and higher remuneration, with no formal procedure for considering the benefits of reducing greenhouse gas emissions. This is why policies are needed to accelerate the expansion of energy supply based on less emissions-intensive sources (Labis, Visande, Pallugna & Caliao, 2011).

Without effective energy policies, locally or ideally a global energy policy, the reduction of carbon intensity in a global energy system will not evolve in a way that is compatible with the goal of reducing emissions worldwide, with effective possibilities of limiting global warming to 2°C by 2020 at the latest.

Energy sectors in many nations, with very few exceptions, are not purely and simply free markets. They are heavily influenced by incentives and regulatory mechanisms such as payments for environmental services, standards, tax exemptions and special financing (Ekins, Pollitt, Summerton & Chewpreecha, 2012).

The special report on renewable energy sources and climate change mitigation issued by the IPCC in 2011 clarifies the context of renewable energy generation in a global effort to mitigate climate change (Pearce, 1991).

The concept of the environmental productivity of energy sources can be adopted as a limiting criterion for the authorisation and granting of incentives for energy sources, with the aim of reducing the intensity of greenhouse gas emissions from energy sources in the energy matrix, and can also function as a criterion for the decommissioning of emission-intensive energy sources above a certain level that can be considered unacceptable progressively over time (Gerber & Opio, n.d).

The "environmental productivity" of energy sources is a defining element of the metric for evaluating energy sources and systems to check their environmental effectiveness, which is understood as the relationship between the energy flows offered for consumption and the amount or extent of the biosphere and atmosphere needed to supply the corresponding energy.

Tests for applying this concept to the Brazilian energy matrix, determining the boundary conditions for a low-carbon energy matrix, a debate on regulatory mechanisms such as energy policy governance instruments and a proposal for necessary changes or new instruments in energy policy can bring the structure of energy production and consumption in Brazil gradually closer to a politically established target of boundary conditions (Ekins, Pollitt, Summerton & Chewpreecha, 2012).

Expanding this study to other energy systems would make it possible to establish a global vision of the potential contribution of a global energy policy based on limitations and incentives linked to minimum levels of environmental productivity of energy sources. It would make it possible to assess the economic and environmental costs and benefits of developing a virtuous cycle of transition in the global energy system. These benefits would result in new and greater investments in renewable energy sources, made possible through regulatory incentives, tax reductions or exemptions and, above all, through financing that could be provided by multilateral funds aligned with a global energy policy, while also using global implementation mechanisms within the framework of the Conference of the Parties (COP) to the Framework Convention on Climate Change (Neimark & Mott, 2011).

The "environmental productivity" of energy sources can be a useful indicator for comparing the effort needed to expand the energy supply environment of a set of available energy sources. It can therefore be used to help define strategies for expanding energy supply from various sources at the lowest environmental effort.

CHAPTER 6

Conclusion

The fundamental relationship between productivity and public policies has profound impacts, as analysed in this study. There is a profound link between energy policies and environmental policies (Vine, 2008).

Considering the first question about how energy is consumed and how the environmental productivity of energy sources can be used to minimise the depletion of atmospheric or biospheric resources when expanding energy generation, and also the question about the supply of energy from different sources and its sustainability and reliability, it can be concluded from the research carried out that:

- Renewable energy sources are the immediate long-term solution in terms of environmental productivity. Expanding their use would increase the environmental productivity of energy sources (Heijman, 1990).
- Adding value is fundamental to achieving sustainability, as is incorporating efficiency measures and increasing productivity in various sectors, including and particularly environmental productivity.

With regard to regulations and legislation that can be adopted to limit the use of energy and the impacts and implications of its production, in order to consider the metrics of the environmental productivity of energy sources, as addressed in the third proposed question; the study allowed us to conclude that:

- There are many public policy reforms in the energy sectors that would be needed to safeguard, regulate and also transform energy use to improve the environmental productivity of the energy matrix.
- The adoption of a metric such as the "environmental productivity of energy sources" can be incorporated into energy policy in a relatively simple way, insofar as the elements necessary for its calculation and use in the decision-making processes for new energy sources, or the decommissioning of existing sources, are based on information that already exists and is incorporated into the respective decision-making processes.

CHAPTER 7

Recommendations

This study recognises that it is crucial to incorporate renewable and non-renewable energy resources into the analysis context. Renewable resources should be given greater emphasis. It is therefore recommended that the prioritisation of and transition to energy sources be stimulated to facilitate the replacement of fossil fuels with renewable energy sources (Wang, 2011). In a real sense, these stimuli would facilitate a reduction in pressure on the environment, especially the biosphere and the atmosphere (Elbehri, Segerstedt & Liu, n.d).

A rapid transition to a predominantly renewable energy matrix is recommended. Society must therefore realise that extreme dependence on fossil energy sources leads to unnecessary pressure on the biosphere and atmosphere (Heijman, 1990). The study's conclusions lead to the realisation that greater energy and environmental productivity results in lower expenses, as well as improving energy security conditions (Elbehri, Segerstedt & Liu, n.d). Bearing in mind the ultimate goal of guaranteeing the ability of consumers in a given context to effectively have lower costs and lower environmental impacts, various measures that can stimulate efficiency in the use of energy are essential for these organisations to have effectively sustainable operations. It is therefore recommended to adopt real energy pricing policies that take into account the real environmental and social costs incurred. Such energy pricing policies would ultimately reduce pressure on the biosphere and atmosphere. The main reason why greater environmental productivity in production and greater efficiency in the use of energy reduce pressures on the atmosphere and the biosphere is that by increasing energy prices and correctly signalling relative prices, based on signals indicating levels of environmental productivity, both consumers and energy market agents will make natural decisions oriented towards greater productivity and efficiency; in this way, the advantage is then seen in putting measures in place that can reduce the cost of energy. In doing so, greater utilisation of renewable energy sources has the best potential for guaranteeing energy productivity for users in general.

It is also recommended that technicians, financiers and financial consultants in this field of study can also act as service providers. Offering their technical and financial consultancy services, providing higher quality energy solutions and

efficiency for small and large users, contributing to their sustainability by reducing operating costs. It is also recommended that these users have access to adequate technical and institutional information, which should be the object of government attention.

Finally, it should be emphasised that there are no policies or their effects, based on incentives for greater environmental productivity and greater efficiency in the use of resources, that can cause negative impacts for energy users. Therefore, experts, technicians and financiers can use this concept to facilitate the formulation and implementation of such policies.

CHAPTER 8

References

Adan, H., & Fuerst, F. (2015). Do energy efficiency measures really reduce household energy consumption? A difference-in-difference analysis. *Energy Efficiency.*

Anjaneyulu, Y., & Manickam, V. (2007). *Environmental impact assessment methodologies.* Hyderabad [India]: BS Publications.

Awange, J., & Kyalo Kiema, J. (2013). Environmental geoinformatics. Berlin: Springer.

Black, B., & Weisel, G. (2010). Global warming. Santa Barbara, Calif.: Greenwood.

Borg, N. (2008). Energy efficiency: past the tipping point? *Energy Efficiency,* 7(1), 77-78. http://dx.doi.org/10.1007/sl2053-008-90Q8-8

Bozorgi, A. (2015). Integrating value and uncertainty in the energy retrofit analysis in real estate investment-next generation of energy efficiency assessment tools. *Energy Efficiency,* 5(5), 1015-1034.

Brown, M., & Herendeen, R. (1996). Embodied energy analysis and EMERGY analysis: a comparative view. *Ecological Economics,* 79(3), 219-235. http://dx.d0i.0rg/l 0.1016/s0921 -8009(96)00046-8

Dinda, S. *Handbook of research on climate change impact on health and environmental sustainability.*

Ec.europa.eu,. *Sustainable Use of Natural Resources - Environment - European Commission.* Retrieved 25 January 2016, from http://ec.europa.eu/environment/natres/studies.htm

Ekins, P., Pollitt, H., Summerton, P., & Chewpreecha, U. (2012). Increasing carbon and material productivity through environmental tax reform. *Energy Policy, 42,* 365-376.

http://dx.doi.org/10.1016/j.enpol.2011.11.094

Elbehri, A., Segerstedt, A., & Liu, P. *Biofuels and the sustainability challenge.*
Environmental comfort and productivity (1987). *Applied Ergonomics,* 81.
http://dx.d0i.0rg/l 0,1016/0003-6870(87)90116-5

Fiala, M., & Bacenetti, J. (2012). Model for the economic, energy and environmental evaluation in biomass productions. *Journal Of Agricultural Engineering, 43(\).* http://dx.doi.org/10.4081/iae.2012.e5

Goldemberg, J. (1994). Science and the climate convention. *Science & Global Security,* 4(3), 407-424. http://dx.doi.org/10.1080/08929889408426409

Goldemberg, J. (1995). Technically a sound case, but not convincing on economics. *Energy For Sustainable Development,* 2(3), 60-61. http://dx.doi.org/10.1016/sQ973- 0826(08)60140-4

Goldemberg, J. (2012). *Energy* (pp. 20-42). Oxford: Oxford University Press.

Hansen, S., & Brown, J. (2004). *Investment grade energy audit.* Lilburn, Ga.: Fairmont Press.

Heijman, W. (1990). *Natural resource depletion and market forms* (pp. 20,30,45,71). Wageningen: Wageningen Agricultural University.

J. Goldenberg, Energy - What everyone needs to know, 2012. *Oxford University Press,* New York, United States

Khayyat, N. *Energy demand in industry* (pp. 20,32,46,54,60).

Labis, P., Visande, R., Pallugna, R., & Caliao, N. (2011). The contribution of renewable distributed generation in mitigating carbon dioxide emissions. *Renewable And Sustainable Energy Reviews , 15(9),* 4891-4896. http://dx.doi.org/10.1016/j .rser.2011.07.064

Lough, T. (1996). Energy analysis of the structures of industrial organisations. *Energy, 21(2),* 131-139. http://dx.d0i.0rg/l0,1016/0360-5442(95)00092-5

M. Wackernagel and W. Rees, Our Ecological Footprint: Reducing Human Impact on the Earth, *New Society Publishers*, Gabriola Island, BC, 1996

McCarthy, J. (2001). *Climate change 2001.* Cambridge, UK: Cambridge University Press.

Metz, B. (2002). Towards an equitable global climate change regime: compatibility with Article 2 of the Climate Change Convention and the link with sustainable development. *Climate Policy,* 2(2-3), 211-230.

Ministry of Agriculture, Livestock and Supply. *National Sugarcane and Agroenergy Balance Sheet,* 2007

Ministry of Science and Technology. Inventory of Anthropogenic Emissions and Removals of Greenhouse Gases not Controlled by the Montreal Protocol - *Initial Communication from Brazil.* 2010

Neimark, P., & Mott, P. (2011). *The environmental debate.* Amenia, NY: Grey House Pub.

Mortimer, N.D., (1991). Energy analysis of renewable energy sources. *Energy Policy, 19(9),* 813. http://dx.doi.org/10,1016/0301-4215(91)90002-6

Publications of the International Atomic Energy Agency (1965). *Nature,* 207(5002), 1135- 1135.

Pinguelli Rosa, et al. Greenhouse gas emissions from hydroelectric reservoirs (2008). Oecol.Bras., 12(1) 116-129, 2008. Rio de Janeiro, Brazil.

R. U. Ayres, L. W. Ayres, and B. Warr, Energy, Power and Work in the US Economy, 2002. INSEAD, Fontainebleau, France.

Recent Energy Studies and Energy Policies in Turkey (2001). *Energy Sources, 23(6),* 495- 510. http://dx.d0i.0rg/l 0,1080/00908310152125148

Silva, Ludimila L. Financial compensation for hydroelectric plants as an instrument of economic, social and environmental development. *University of Brasília,* 2007.

Simón, X. (2012). Eolic energy and rural development: an analysis for Galicia. *Spanish Journal Of Rural Development,* 13-28. http://dx.doi.org/10.5261/2012.gen3.Q2

Smil, V. (2008). *Energy in nature and society.* Cambridge, Mass.: MIT Press.

Smithson, P. (1989). Changes in Global Climate. *Quaternary Science Reviews, 8(2),* 203.

Torrico Albino, J., & Janssens, M. (2011). ENVIRONMENTAL IMPACT ASSESSMENT OF LAND USE SYSTEMS USING EMERGY IN TERESÓPOLIS-BRAZIL. *J. Nat. Resour. Dev.* http://dx.doi.org/10.5027/jnrd.v2i0.02

Vine, E. (2008). Strategies and policies for improving energy efficiency programmes: Closing the loop between evaluation and implementation. *Energy Policy, 36(10),* 3872-3881.

W. Chern, W. James. Measurement of Energy Productivity in Asian Countries, *The Ohio State University,* USA

Wang, C. (2011). Sources of energy productivity growth and its distribution dynamics in BraziL *Resource And Energy Economics,* 55(1), 279-292. http://dx.doi.org/10.1016Zj.reseneeco.2010.06.005

Williams, R., McKane, A., Perry, W., Aixian, L., & Tienan, L. (2005). *Linking Energy Efficiency and ISO.* Washington, D.C.: United States. Dept. of Energy.

Yan, H. *The Eco-Social Impacts of Biofuel Expansion in Brazilian Amazon | Amazonian Biofuels. Amazonian Biofuels.* Retrieved 25 January 2016, from http://sites.duke.edu/amazonianbiofuels/biofuels-effects-on-the-environrnent- overvíew/

Zubrin, R. (2007). *Energy victory.* Amherst, N.Y.: Prometheus Books.

CHAPTER 9

Annexes

9.1. Annex I - Analysis of energy efficiency and efficiency programmes

Some examples of industrial consumer classes that generally use energy intensively as an input were chosen in order to, based on an analysis of their consumption, both on a small and large scale, understand how the environmental productivity of energy sources can be established as a benchmark indicator for their sustainability programmes, in particular results of an energy audit were analysed, including research into energy conservation and productivity enhancement measures, and a synthesis framework of analysis (Bozorgi, 2015). This was done by means of a feasibility and management study, additionally including data on the availability of equipment for efficient energy utilisation by energy users (Hansen & Brown, 2004). The results of the analysis on energy use are presented, as well as possibilities for introducing the environmental productivity indicator for energy sources into energy procurement decisions and sustainability programmes for energy users.

Bearing in mind the subsidiary purpose of this thesis, we accept that this analysis was based on energy audit reports produced in other works. This type of study can also be carried out by analysing the economy and efficiency of energy consumption ("Environmental comfort and productivity", 1987). Here, the analysis focused on three methods for increasing energy efficiency: (i) a roadmap for efficient consumption behaviour, (ii) energy transformation and changes to more efficient forms, and (iii) innovations.

Improvements in the management of energy installations, under the responsibility of energy users, also imply lower costs and effective mechanisms for increasing energy efficiency. Programmes to analyse energy demands and energy consumption efficiency in facilities and equipment, generally conducted through energy audits, are effective in identifying potential savings and productivity increases.

Mechanisms for increasing energy productivity, conceivable with new technologies, increase the chances of reducing energy consumption.

The energy efficiency programmes considered are presented below:

Energy audits

An energy audit is a study of a client's facilities in which energy is used. The study involves auditing the facilities and the use of energy for their operation, making it clear to the client how energy is used and the practices adopted in this use, presenting the effects that can be achieved with energy efficiency projects. This type of energy audit raises awareness through instructions on how to use energy better and urges users to adopt the auditors' proposals (Hansen & Brown, 2004).

Discount Programme

It consists of two segments: one in which the energy user receives cash or rebates, and a second in which the manufacturer of more efficient equipment receives refunds. In the cash or rebate programme, the customer is entitled to the benefit whenever they buy certain more efficient equipment that qualifies under the programme. In the rebate programme, the manufacturer of more efficient equipment (qualified in the programme) is reimbursed for expenses incurred in manufacturing that reduce the cost of its products for its customers. This helps users decide to purchase more efficient equipment (Awange & Kyalo Kiema, 2013).

Direct installation programme

In this programme, the energy supplier has projects aimed at introducing efficient equipment to its customers. One example is the replacement of lighting systems with others with lower specific consumption (such as LED lighting systems). In this type of programme, the motivation for the utility's direct investment is to free up capacity in the electricity system to serve other loads at a lower cost than investing in expanding the electricity system's generation, transmission and distribution capacity. The benefits of reducing costs in the electricity system are measurable.

Education and training programme

This type of programme promotes the education and training of all agents involved in the energy supply and use chain, for example marketers, engineers, workers and facility assessors, on energy saving, equipment and measures that result in lower consumption, as well as points that should be monitored.

Loans and financing or grants

Initial costs, such as improving installations and replacing equipment, are disincentives. Loan and financing programmes aim to provide credit to users and make it possible to cover expenses incurred at the start of implementing energy efficiency plans (Simón, 2012).

Standard performance bids / contracts

This type of programme aims to outsource the implementation of energy efficiency measures through performance contracts with energy efficiency specialists to supervise and update energy efficiency and productivity programmes for end users. Specialists in making energy use more efficient and production more productive are better placed than electricity distribution companies to identify opportunities for improvement, maximising the productivity of energy users.

Upstream incentives

This type of programme aims to bring motivators or collaborators together with manufacturers, traders or speculators in the energy market to incorporate energy efficiency projects. These projects help to expand investment in the use of efficient energy (Williams, Perry, Aixian & Tienan, 2005).

Failure replacement programme

This type of programme aims to promote the introduction of more efficient equipment to replace existing equipment in the event of a failure. This practice can be best implemented with tradesmen or providers of equipment replacement services in companies, as in the case of heating, ventilation and air conditioning equipment maintenance services.

Early replacement programme

This is aimed at replacing equipment that is in operation, not sufficiently faulty, but that justifies its replacement. Replacement with more efficient equipment is justified financially. In this case, only arguments and extraordinary gains are justified, particularly when the reduction in operating costs is evident.

New facilities programme

When there is a new paradigm for the form of production or the organisation of systems that involve the use of energy, there is usually an opportunity to introduce more efficient and productive energy practices. In general, the design of a process begins when it is first conceived and involves an initial cost corresponding to that conception. Subsequent changes represent costs, and are generally only implemented in specific and limited aspects. It is therefore in the design of a new process that new arrangements and more efficient equipment can be introduced.

Commissioning

The success of energy efficiency and productivity programmes is linked to a change in paradigms and culture on the part of energy users. It is important to mark and recognise when these changes occur and materialise with the implementation of projects involving greater efficiency and productivity in energy use.

One example is the adoption of events to publicise improvements in production processes and new products or services that are offered to end users with lower energy consumption or less environmental impact.

With this type of programme in mind, manufacturers seek to create expectations among their customers about new and more efficient products (Ueda et al, 2009).

It can be seen from the business media that there is a growing tendency for companies to consider the need to invest in specialised media strategies in order to increase the esteem in which their products are held by end users (Broil et al., 1994). This is a process of interaction between manufacturers and users. Users increasingly demand higher quality and meaning from products, and manufacturers understand this demand and introduce innovations and higher quality, competing with each other in the market, and creating a virtuous environment in that increased productivity occurs as a consequence of this process (Heikki, 2000).

9.2. Annex II - Economic evaluation of energy efficiency implementation

Ensuring energy efficiency entails some costs, such as the cost of purchasing equipment, installation and maintenance expenses. Companies and other end users of energy become motivated to implement energy efficiency measures if the financial results offset the investment and implementation costs. In this assessment, the concept of net present value (NPV) is used. The net present value is characterised by the difference in the sums at present value between the projected costs with and without the implementation of the energy efficiency project in question. This analysis is useful for detailing the real benefits of implementing a project. In the case of energy efficiency projects, their justification ends up directly corresponding to the increase in productivity resulting from their implementation. The assessment of net present value in a scientific manner (Lim et al, 2008) can be developed based on the following equations.

$$I = p \times (L \times P \times 365 \times 24) \quad (7)$$

Where: I = Annual project revenue

p = Unit price of electricity in kWh

L = Plant utilisation factor

P = Installation power in kW

e

$$E = E_f \times P + E_v \times Q \quad (8)$$

Where: E = Annual operating and maintenance expenses

E_f = Fixed operating and maintenance cost in kW/year

E_v = Variable operating and maintenance cost in kWh

P = Installation power in kW

Q = Amount of energy made available per year in kWh

Another way of assessing the financial viability of an energy project's productivity is to evaluate the pay-back time, which is a more direct form of assessment. The pay-back time is the measure of the time needed to recover the financial resources invested in the project's implementation. The general issue is that this method reduces the analysis to financial speculation. The mathematical sentence used to analyse a project's payback time is as follows:

Payback period = Investment required / net cash flow (9)

The payback period is expressed in years and in general the shorter the payback periods the more advantageous the projects become. On the other hand, short payback periods are not the most satisfactory measures for assessing the long-term benefits of energy efficiency and productivity projects (Honey bee, 2008). This payback period system can fail to indicate the real benefit of a project if it is not analysed more thoroughly.

Most companies consider a payback period of 0 - 3 years as the limit for resuming investment in the implementation of a project, but in a study carried out on the German market in 1991, it was found that the payback periods required were generally 4.1 years (Gruber and Brand, 1991).

It is clear that the decision on whether to implement an energy efficiency project, both on the demand side and on the energy production side, can be significantly improved if additional indicators are introduced to add understanding of implications that are generally not expressed in essentially financial methods of analysis, notably indicators that express the extent of the real costs associated with decisions to continue using energy as in conventional arrangements, compared to alternatives from other sources and forms of energy use that are more efficient and have a lower environmental impact.

9.3Annex III - Cost reduction of photovoltaic technology

Although it is still the most expensive electricity generation technology available commercially, it has seen an accelerated reduction in costs, as shown in Graph 4.

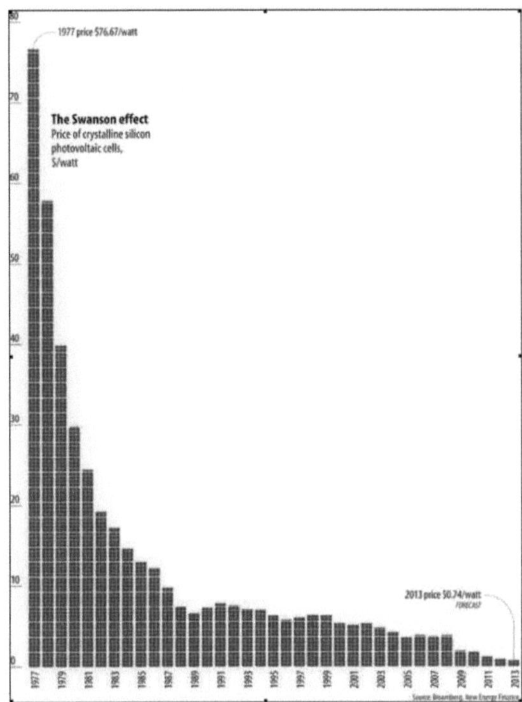

Graph 4. Price reductions for polycrystalline silicon photovoltaic cells.

Source: International Energy Agency, 2012

The price of photovoltaic modules has fallen by more than 70 per cent over the last decade, resulting in a significant reduction in the cost of generating electricity in photovoltaic plants to levels of between 100 USD/MWh and 260 USD/MWh. This cost level at its lower limit already allows for the replacement of some more carbon-intensive conventional energy sources (Ekins, Pollitt, Summerton & Chewpreecha, 2012).

Printed by Books on Demand GmbH, Norderstedt / Germany